Black Holes

AUDREY E. RANDLES

Copyright © 2012 Audrey Elizabeth Randles
All rights reserved.

Preface

The Theory of Matrix series of books offers the exiting developments in cosmological theory. We combine elements of psychology, cosmology, and astrophysics to discover secrets hidden deep in the Universe.

'Black Holes' is the 5th book of the series. In this book, we summarise the general part of the Theory of Matrix related to the black holes and the multimodal systems holding them.

The Universe contains an infinity of the Matrixes, 'temporal fields', creating streams and rivers of time. Space sometimes is looking flat and time moves the world.

Masses, associated with the visible space, are rotated by the centrifugal force inside the temporal Space-Time Matrixes, around their mass centres, at 'o' time-points of our understanding, like the surface of water rotating in a bucket. Space-Time flux changes properties and builds channels, path shorting and bringing forth spaces close together or further from us.

Stay well, and enjoy your reading.

Yours sincerely,
Audrey E. Randles

Contents

Preface
Acknowledgement
Introduction
Chapter 1 Types of Black holes
Chapter 2 Energy Curves Space-Time
Chapter 3 Space-Time and Energy Imbalance
Chapter 4 The Equivalence principle
Chapter 5 Central Transmitting Black Holes
Chapter 6 Surface-orientated Black Holes
Chapter 7 Binary Systems of Black Holes
Chapter 8 Gravity of the Black Holes
Chapter 9 Reverse of the Absorbing Systems
Chapter 10 Reverse of the Black Holes
Chapter 11 Antigravity of the Black Holes
Chapter 12 Rotation of the Black Holes
Afterword
Content Use Policy

Acknowledgement

We would like to express our gratitude to the National Aeronautics and Space Administration (NASA), NASA's Goddard Space Flight Center (GSFC), NASA's Jet Propulsion Laboratory (JPL), the California Institute of Technology (Caltech), the Infrared Processing and Analysis Center (IPAC), Event Horizon Telescope Collaboration, and the University of California at Los Angeles (UCLA) for the excellent images and exciting descriptions, published by NASA's Jet Propulsion Laboratory (JPL) and used in this book to illustrate the spectacular beauty of our dynamic world, its complex structures, and success of the space explorations addressing fundamental questions of the Space physics and human psychology concerning our place in the Universe.

The views and opinions of the author, expressed in this book, do not necessarily state or reflect those of NASA's Goddard Space Flight Center, NASA's Jet Propulsion Laboratory, the California Institute of Technology, the Infrared Processing and Analysis Center, Event Horizon Telescope Collaboration, the University of California at Los Angeles or National Aeronautics and Space Administration.

Image 1: 'Coiled creature of the night' Image Credit: NASA/JPL-Caltech

'The "eye" at the center of the galaxy, called NGC 1097, is actually a monstrous black hole surrounded by a ring of stars. The black hole is huge, about 100 million times the mass of our sun.' NASA

Introduction

The concepts of John Michell and Pierre-Simon Laplace, related to the massive celestial bodies with the intense gravity that light could not escape, continue being actual from the 18th century until nowadays.

The existence of the black holes can only be detected by observing their gravitational interactions with the direct surroundings. The modern interpretations, multiple theorems and equations describe the different hypothetical and computer-simulated physical properties of black holes. Questions associated with the black hole information loss paradox, gravitational time dilation, space, mass and energy within black holes remain.

Black holes do not directly emit any signals. Nasa scientists study black holes effects on the matter nearby.

Researches learn about the properties which the transmitting black holes display in our dynamic world. They learn about the external accretion disks of black holes. These disks are formed from the dust and gas as a galaxy or a star, attracted by the black hole's gravity and fallen onto the black hole.

They learn about the quasars, formed as the dust of the accretion disk heating up by friction, and jets of radio waves and other forms of background radiation; they learn blueshift and redshift effects associated with the black holes.

They learn about the ergosphere of the black holes and the 'photon sphere' that is a spherical 2-dimensional boundary in which photons that move on tangents to that sphere would be trapped in a circular orbit about the black hole. Light, crossing the photon sphere, will be processed by the black hole. In this book, we will discuss the 2-dimensional structures of background radiation and its associations with the black holes.

Image 2: 'Black hole grabs starry snack (Artist Concept)' Image credit: NASA/JPL-Caltech

'This artist's concept shows a supermassive black hole at the center of a remote galaxy digesting the remnants of a star. NASA's Galaxy Evolution Explorer had a "ringside" seat for this feeding frenzy, using its ultraviolet eyes to study the process from beginning to end. The artist's concept chronicles the star being ripped apart and swallowed by the cosmic beast over time.

'First, the intact sun-like star (left) ventures too close to the black hole, and its own self-gravity is overwhelmed by the black hole's gravity. The star then stretches apart (middle yellow blob) and eventually breaks into stellar crumbs, some of which swirl into the black hole (cloudy ring at right). This doomed material heats up and radiates light, including ultraviolet light, before disappearing forever into the black hole. The Galaxy Evolution Explorer was able to watch this process unfold by observing changes in ultraviolet light.' NASA

The researchers learn about an event horizon. In this book, we will provide the Theory of Matrix clarification on the matter.

Different models for the early Universe and Big Bang theory cannot find the answer to the main questions - what allowed the primordial black holes to form, where the matter came from to become the uniform mass or any other dense medium, and where the initial density perturbations came from to grow under their gravity.

Besides, most of the modern theories state that the fabric of space is damaged in the black holes and energy and information are lost.

The theories involving singularities, such as Space-Time singularity, ring singularity, and gravitational singularity with the infinite Space-Time curvature, the infinite singular region with zero volume and infinite density, and 'spaghettification' or the 'noodle effect', do not provide us with the satisfactory answers.

The theories of singularity try to put the body of the existing Universe into the point - a singularity with the utmost infinite mass, and explain its birth in terms of a Big

Bang as a necessary concept for the beginning of the Universe.

Let us compare the birth of our Universe with childbirth. You do not need to suppress the mass of the adult person, his life history and life experience into one point to make him fit into the mother's womb to explain a natural process of childbirth and development.

Similarly, you do not need to suppress the body of the existing Universe, its mass and history into a singularity and explain the birth of the Universe as a result of a Big Bang. The Big Bang Theory has two weak points - the beginning of the Universe existence from nothing and the end of the Universe existence to nothing. The Law of the Energy Conservation is violated. Nothing is coming from nothing and going to nothing.

The theory of singularity is still left with the same problem - the start of the singularity. The reverse of time in the frame of the theory of singularity will not lead us to the correct decisions associated with gravity and antigravity. The theories of singularity and Big Bang reflect limitations of modern mathematics.

There is no need for the Universe beginning from a Big Bang and the Universe final collapse. Space-Time is complete and harmonic, and characteristics of the Universe are different in different modalities of Space-Time.

The researchers learn about the event horizon of a black hole as a boundary in Space-Time through which matter, volumes, energy and information pass towards the black hole, its topology, and the effect, known as gravitational time dilation at the event horizon. To a stationary observer, an object, falling into a black hole, appears to slow down as it approaches the event horizon, and the clock near a black hole would appear to slow down the time compared with the time of the stationary observer and the stationary clock located far away from the black hole. Simultaneously, the clock of the observers, falling into a black hole, would appear to them to tick as usual.

The researchers learn about the rotation of black holes, neutron stars, pulsars, quasars, and blazars. You will see some of the impressive images and NASA's exiting descriptions chosen for this book. We describe transmitting black holes, their structure and mechanics, and offer the conceptual answers to the questions mentioned above.

Image 3: 'Spitzer observes neutron star collision' Image credit: NASA/JPL-Caltech

'NASA's Spitzer Space Telescope has provisionally detected the faint afterglow of the explosive merger of two neutron stars in the galaxy NGC 4993. The event, labeled GW170817, was initially detected in gravitational waves and gamma rays. Subsequent observations by dozens of telescopes have monitored its afterglow across the entire spectrum of light. The event is located about 130 million light-years from Earth.' NASA

Types of Black Holes

The galaxies with supermassive black hole candidates include the Andromeda Galaxy, Sombrero Galaxy and our Milky Way. The supermassive black holes have been found at the centres of most galaxies.

According to the Theory of Matrix, there are three main types of black holes: non-transmitting black holes, transmitting non-radiating black holes and transmitting radiating black holes.

A non-transmitting black hole is an enclosed hollow within the Space-Time region of the dynamic body of the Universe that needs to be filled with space, time, and energy.

A transmitting black hole is a flexible Space-Time cordlike structure joining two or more unbalanced objects or systems together and, accordingly, building the system with two or more centres. We can observe two or more galaxies with the black holes building a remote galaxy or a black hole in association with a galaxy or a star.

A transmitting black hole is a contact path equilibrating the system. It is a subspace whose points or elements are all in another space. It is a flexible Space-Time cordlike structure attaching the system, holding a black hole, to other Space-Time regions.

A transmitting non-radiating, or absorbing, black hole transmits volumes, matter, and energy from outer space into the central areas of the unbalanced system holding the black hole. The system, usually the unbalanced non-radiating system, utilizes the volumes, mass, and energy delivered via the black hole. A transmitting non-radiating black hole is a flexible Space-Time structure, formed by the energy of the unbalanced non-radiating system and arranged by background radiation. This flexible structure is attached to the unbalanced Space-Rising system, similar to an umbilical cord attached to the 'placenta' during gestation. It is usually associated with the centre of the unbalanced absorbing Space-Rising system, which is sometimes undetectable in our dynamic world.

The properties of the 'gravitating' black hole are related to the Space-Time and Mass-Energy imbalance, reflected in the intense gravity at the centre of the unbalanced Space-Rising system. A black hole's 'opening' into the central areas of the maternal system is associated with the Space-Time and Info-Energy imbalance reflected in the intense gravity at the centre of the system. A transmitting non-radiating, or absorbing, black hole 'gravitates' in the outer space. A black hole 'gravitates' if the central region of the maternal system displays the deficit of volumes, mass, kinetic energy, and other energy representations.

A black hole 'radiates' if the central region of the maternal system displays the excessive volumes, mass, kinetic energy, and other energy representations. The properties of the transmitting radiating black holes are related to the Space-Time, Info-Energy, and Mass-Energy imbalance, reflected in the intense antigravity at the centre of the maternal system.

A black hole 'evaporates' if the Space-Time, Info-Energy, and Mass-Energy of the associated system are balanced. The dynamic balance within the system, holding a black hole, happens when a process and its reverse are occurring at equal rates.

NASA researchers describe black holes as follows,

'Black holes are tremendous objects whose immense gravity can distort and twist space-time, the fabric that shapes our universe. These effects, consequences of Einstein's general theory of relativity, result in the bending of light as it travels through space-time. By looking for these light distortions in X-rays streaming off material near black holes, researchers can gain information about their spin rates... The faster a black hole spins, the closer its accretion disk can lie to it - another consequence of Einstein's theory of relativity.' NASA

Black holes of the early universe

Image 4: 'Prehistoric black hole (Artist's Concept)' Image credit: NASA/JPL-Caltech

'This artist's conception illustrates one of the most primitive supermassive black holes known at the core of a young, star-rich galaxy. Astronomers using NASA's Spitzer Space Telescope have uncovered two of these early objects, dating back to about 13 billion years ago...

'The monstrous black holes are among the most distant known, and appear to be in the very earliest stages of formation, earlier than any observed so far. Unlike all other supermassive black holes probed to date, they lack dust... the dust tori are missing and only gas disks are observed. This is because the early universe was clean as a whistle. Enough time had not passed for molecules to clump together into dust particles.

'Some black holes forming in this era thus started out lacking dust. As they grew, gobbling up more and more mass, they are thought to have accumulated dusty rings.

'This illustration also shows how supermassive black holes can distort space and light around them (see warped stars behind black holes). Stars from the galaxy can be seen sprinkled throughout, and distant mergers between other galaxies are illustrated in the background'. NASA

Energy Curves Space-Time

We mentioned above that the transmitting black holes join two or more unbalanced systems together and the properties of the black holes reflect the Space-Time, Info-Energy, and Mass-Energy imbalance at the centre of the maternal system. Accordingly, to consider the functioning of the black holes, we need to make clear the properties of the associated systems.

Every existing single object and every system of the objects, including visually 'empty' spaces, subatomic particles and holes, stars and planets, galaxies and star clusters, systems holding black holes, and the thermodynamic Universe, are the multimodal objects and systems. The multimodal objects and systems carry the complex properties of multiple Space-Time modalities. Characteristics of the objects and systems are various in different modalities of Space-Time.

The 1-dimensional, 2-dimensional, and 3-dimensional elements of a system are integrated into the system's multimodal Space-Time and Info-Energy structure. We call the multimodal Space-Time and Info-Energy structure of a system - the Matrix for short. The Matrix depicts the total mass and energy of a system in space and time, including its past, present, and future.

According to the theory of General Relativity, energy curves Space-Time. Please see Einstein Albert, 'Relativity: The Special and General Theory' (1916). We support this idea.

The system's Matrix is shaped by the regular repeated (with the equal distances, typically rectangular) grid-like 2-dimensional arrangement of the energetic net-structure. The infinitely thin filament forms the Matrix of every existing system. The 2-dimensional grid contains the system's potential Info-Energy.

We use the term 'Info-Energy' (forms of energy) to describe the energy and information, which are complementary and cannot be separated. Energy and information are the fundamental characteristics of the Universe and the existing objects and systems. Info-Energy is to be defined in relation to a frame of reference. Mass-Energy is 'the mass of a body regarded as energy, according to the laws of relativity' (Oxford Dictionary). We, respecting the laws of relativity, use the term 'Mass-Energy' as interchangeable with the term 'Info-Energy' of an object or a system.

The total Info-Energy of the Matrix grid is the total potential energy and associated potential information, which the existing object or the system possesses in space and time, including its past, present, and future. Background radiation, including those known as cosmic background radiation, and light arrange the Info-Energy of the Matrix grid of the existing objects and systems.

The grid composes the fundamental 2-dimensional Space-Time of the multimodal system. The potential space of the Current Time forms the 2-dimensional 'container' for the system's volume filled with Info-Energy acting in the volume of the system in the period of time now occurring. The Matrix grid limits the system's volume at the centre of the Matrix and shapes the system's Spaces of Time. The Matrix grid has a form of the Riemannian Manifold with the positive or negative curvature, depending on the properties of the system. Riemannian geometry applies to its investigation.

Two types of the Matrixes have been identified - the Time-Rising Matrix (TRM) and the Space-Rising Matrix (SRM).

The TRMs are the property of the radiating objects and systems, including 'black body' radiation spectrum. The 2-dimensional grid of the TRM has a form of the Riemannian Manifold with the negative curvature (Figure 1).

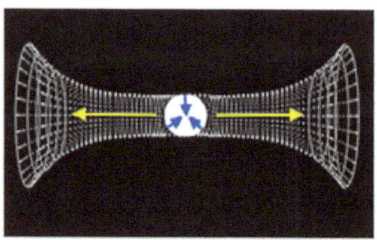

Figure 1: Time-Rising Matrix (TRM)

The SRMs are the property of the non-radiating objects and systems. The 2-dimensional grid of the SRM has a form of the Riemannian Manifold with the positive curvature (Figure 2).

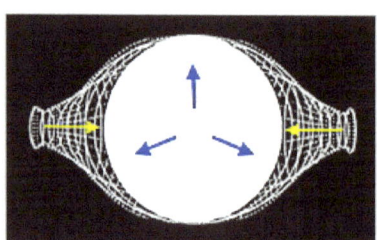

Figure 2: Space-Rising Matrix (SRM)

Strictly speaking, the reproduction of the Matrix 2-dimensional grid on the paper is not exact. It is related to the difficulties to reproduce the 2-dimensional structure on the paper.

It is a theoretical mistake that one can currently reproduce a 2-dimensional structure on the paper. A drawing on the paper and PC screen does not reflect a 2-dimensional structure - our world-perception influences on making drawings and PC programming. The relative difference between our perception of space and time makes this world of perceptible forms being sensible to us. Accordingly, the structures, which are reproduced on the paper or PC screen, regardless of how special they would be, are no more than our projections of the 3-dimensional systems. Their proportions and structure do not reflect characteristics of the 2-dimensional structures. Similarly, proportions and directions of the 2-dimensional structure in the multimodal Space-Time are entirely different from our projection of any structure on the surface.

Besides, one cannot create a 2-dimensional structure using 3-dimensional particles, one layer of 3-dimensional cells, crystals or atoms, even with the help of nanotechnology. The closest object to the understanding of the 2-dimensional space structure is an ideal square wave.

The exact reproduction of the Matrixes (Figures 1-9) without damaging the main idea is impossible on the paper or computer simulation. Although the shapes of the identified types of Matrixes are correct, we rely entirely on the description.

The Arrows of Time are drawn as the yellow arrows on the reproduction of the Matrixes, while the direction of the Space of the Current Time - the actual space of the period of time now occurring, is represented by the blue arrows, situated following the human perception of the objects and systems as exhibiting three space dimensions (x, y, and z, or a combination of three directions, which can be chosen from the terms: length, width, height, depth, and breadth).

In the General Theory of Relativity, Albert Einstein described gravity as a consequence of the curvature of Space-Time caused by the uneven distribution of mass. In astrophysics, an event horizon is a boundary beyond which events cannot affect an observer. In our dynamic world, we operate in the Space of the Current Time in accordance with our perception of time as a point 'now'. As we sum these statements up, we can introduce the boundary of the curvature of the system's Space-Time as an event horizon that surrounds the total volume of the system that equals the system's Space of the Current Time. As such, the external boundary of the curvature of the Earth Space-Time - affected by the Earth gravity, is the external boundary of the Earth Space of the Current Time. In case of a high energy massive radiating system, such as the Earth, the system's Arrows of Space start at the external boundary of the curvature of the system's Space-Time affected by gravity. In our descriptions, we refer the boundary of the curvature of the system's Space of the Current Time as 'the surface of the system'.

The Space of the Progressive Time is the space of the Matrix cone related to the future of the system (Figures 1-9 - the left cone). The Space of the Regressive Time is the space of the Matrix cone associated with the past of the system (Figures 1-9 - the right cone).

According to Albert Einstein, 'the ponderable masses will be the determining factor in producing the field, or, according to the fundamental result of the Special Theory of Relativity, the energy density...' [Albert Einstein, A Brief Outline of the Development of the Theory of Relativity

(1921)]. Under the Theory of Matrix, the system's multimodal structure is not the field but the energy structure. The ponderable masses are the determining factors in producing the energy density, information density, and multimodal Space-Time and Info-Energy structures of the existing systems.

The space-time-energy Continuum fills up spaces of our past and the future. The 1-dimensional space-time-energy is a property of every existing system. Please see more details about the Continuum in my book 'Space and Time'.

The system's Space of the Current Time at the Matrix centre is filled with the actual Info-Energy. We can directly perceive and measure the systems' volumes and the forms of actual energy, such as masses, kinetic energy, and other forms of energy acting in the volumes of the systems existing in our dynamic world in the period of time now occurring. The complete actual information is coded and fixed by the system's actual energy and represented as the forms of energy and matter in the volume of the system. The density of information is proportional to the density of energy and matter associated with the information.

The Space of the Current Time equals the volume of the system embedded at the centre of the Matrix. The volume of the system is filled with the total actual Info-Energy, acting in the volume of the system in the period of time now occurring. The complete actual information is coded and fixed by the total actual energy in the volume of the system.

The system's total actual Info-Energy is an integrated form of energy, actually acting in the volume of the system in the period of time now occurring and represented in mass, kinetic energy and other energy representations.

The space-associated latent 2-dimensional Info-Energy of the system's grid forms the potential space of the period of time now occurring. The system's potential space underlies the system's volume within the multimodal Space-Time and Info-Energy structure.

The time-associated 2-dimensional Info-Energy of the system's grid forms the time-limits of the system's existence. It forms the Spaces of Time. Energy, building time, is latent Info-Energy associated with the past, present, and future of the system.

The system's Matrix grid is the 2-dimensional framework of the system. It holds a complete set of data related to the embedded system, similar to a chromosome of a living cell. It carries information in the form of Info-Energy blocks.

The notion of the curvature of the 2-dimensional grid is to be defined in a way that is intrinsic to the manifold. The curvature depends on the Space-Time, Info-Energy, and Mass-Energy properties of the system. The curvature of the Matrix grid does not depend on how the surface is enclosed in 1-dimensional, 3-dimensional or higher-dimensional Space-Time. The curvature of the Matrix does not depend on how the system's volume is inserted at the Matrix centre.

Black pixels, reading zero and corresponding to the latent information, are fixed by the latent energy of the 2-dimensional grid. They are associated with the timing mechanisms and sweep rates, and the address of a pixel corresponds to its Space-Time coordinates.

We suppose that the blocks of the data, involved in the actuality or located relatively close to the point 'now' in time, are situated closer to the centre of the Matrix, than other latent information. Hypothetically, if the Info-Energy blocks are associated with the Matrix Space of the Current Time, then they might have a 3-dimensional structure. The distances along the paths and angles are to be measured as the characteristics of the system's latent, or potential, Info-Energy.

The latent energy of the 2-dimensional grid is arranged by the 2-dimensional representations of background radiation, including those currently known as cosmic background radiation, being dynamic in our dynamic world. Accordingly, the Matrix 2-dimensional grid, being the external framework of the system in the multimodal Space-Time, is simultaneously the dynamic internal framework and Space-Time and Info-Energy skeleton of this system in our dynamic world. An example of the Matrix grid influence on the macro-scale is the subatomic and atomic processes associated with the regulation of heat in the body of the Universe, and the example of the grid-forming energy is the energy structure built by the cosmic microwave background radiation carrying this heat and information associated with the regulation of heat in the Universe and objects and systems existing in the Universe.

Consequently, every existing object and every system of the objects is a multimodal structure carrying the properties of different Space-Time modalities. The Matrixes depict the total Space-Time and Info-Energy of the existing multimodal objects and systems.

The total Info-Energy of the Matrix is the total energy and associated information that the existing system possesses in space and time, including its past, present and future.

The systems of the objects of limited mass and energy are limited in space and time. Accordingly, the Matrixes for the systems of the objects of limited mass and energy are limited in space and time.

The potential and actual Info-Energy structures counteract and keep a balance at the central point of the Matrix symmetry, and on the Space-Time axis under the Theory of Matrix natural laws.

It seems reasonable measuring the properties of the Matrix centre and the current characteristics of a system in the standard units related to the current understanding of space, time, energy, and force. Different measurement systems must be applied to the currently inaccessible Spaces of the Progressive and Regressive Time and the 2-dimensional Info-Energy grid that forms them.

Image 5: 'Monsters in space (Artist's Concept)' Image credit: NASA/JPL-Caltech

'This artist's concept illustrates a supermassive black hole with millions to billions times the mass of our sun. Supermassive black holes are enormously dense objects buried at the hearts of galaxies. (Smaller black holes also exist throughout galaxies.) In this illustration, the supermassive black hole at the center is surrounded by *matter flowing onto the black hole in what is termed an accretion disk. This disk forms as the dust and gas in the galaxy falls onto the hole, attracted by its gravity.*

'Also shown is an outflowing jet of energetic particles, believed to be powered by the black hole's spin. The regions near black holes contain compact sources of high energy X-ray radiation thought, in some scenarios, to originate from the base of these jets. This high energy X-radiation lights up the disk, which reflects it, making the disk a source of X-rays. The reflected light enables astronomers to see how fast matter is swirling in the inner region of the disk, and ultimately to measure the black hole's spin rate.' NASA

Space-Time and Energy Imbalance

There is no 'empty' space in the Universe, and no isolated objects and systems exist in our dynamic world. The 'outer space' is always another object or a system. Objects and systems, existing in the Universe, are tangled together by the Space-Time, Info-Energy, and Mass-Energy imbalance.

The unbalanced Time-Rising systems, such as high energy massive radiating systems, including planets, stars, radiating galaxies and star clusters tangled by gravity, undergo their natural development by increasing in time.

The unbalanced Space-Rising non-radiating systems, such as systems holding black holes, self-rising vacuum, and our Universe, undergo their natural development by increasing in size. Visually 'empty' spaces are distinct but weak non-radiating systems. The widening, self-rising quantum vacuum is the strong Space-Rising system. Despite a likelihood of being filled with the dark matter and dark energy, the widening, self-rising vacuum would contain the central transmitting region in the second time dimension facing our dynamic world.

The Space-Time imbalance and the associated Info-Energy and Mass-Energy imbalance are detectable in the second time dimension and 2-dimensional time settings. Please see my book 'Space and Time' for details. Besides, applying the second time dimension to the unbalanced, multimodal structure of a system, we detect a new centre of the Matrix symmetry. Two different representations of the '0' time-point, existing in 2-dimensional time settings within the system's Matrix, indicate the system's Space-Time and Info-Energy imbalance. The unbalanced Matrixes display the curved line of the structured representation of the unbalanced function that is specific to the related system. The Space-Time, Info-Energy, Arrows of Time, and Arrows of Space are not harmonically balanced in the Matrixes of these systems.

Gravity

The Space-Time imbalance, supported by Info-Energy imbalanced, such as the excessive time and time-associated potential Info-Energy and deficit of space, mass, and kinetic energy, are reflected in gravity in the peripheral regions of the high energy massive radiating systems and at the centres of the Space-Rising non-radiating system. This Space-Time imbalance and the associated Mass-Energy and Info-Energy imbalance we perceive as gravity on the surface of high energy massive radiating systems. For example, we can perceive gravity on the surface of the Earth and detect signs of gravity in other high energy massive radiating systems, such as stars and planets. In modern physics, the deficit of mass and kinetic energy of the system is called 'the negative gravitational potential energy'. The Space-Time imbalance and associated Mass-Energy and Info-Energy imbalance at the centres of the spacious, Space-Rising and toroidal non-radiating systems, frequently holding the central transmitting black holes, are sometimes referred in modern physics as 'the zero-point radiation of the vacuum' of Feynman and Wheeler. Nasa scientists observe and photograph black holes by detecting their effect on the matter nearby.

Antigravity

Processes, reflected in antigravity in the Universe, impact on human concepts of 'limitless' or 'endless' in space, extent, and size; mathematical concepts of transfinite and infinite numbers, and ideas corresponding to the infinite Universe and the infinite number of stars in the Universe that is 'impossible to measure or calculate'. It reminds me of the time before Leif Erikson discovered 'Vinland'.

The Space-Time imbalance, supported by Info-Energy imbalanced, such as the deficit of time and time-associated potential Info-Energy and excessive space, mass, and kinetic energy, are reflected in antigravity at the centres of the high energy massive radiating systems and in the peripheral regions of the Space-Rising non-radiating system. In modern physics, the excessive mass and kinetic energy of the system is called 'the positive mass-energy'.

The traditional understanding and management of gravitation did not lead us to antigravity exploration. Antigravity existence is rejected as 'impossible'. The fact of the antigravity existence must not be rejected. Anti-gravitational deceleration and associated processes must be investigated. Antigravity, prompted by the centre of the Earth, balances gravity prompted by the peripheral region of the Earth and therefore protects the planet from the gravitational shock.

The observer, who could reach the centre of the Earth, would be stopped by the 'natural limit' - the impossibility of moving further to the centre and more profound.

Should we travel through the radiating globular star cluster towards the centre, we would experience resistance of the 'empty' space on the way as a result of space and matter deflation and concentration, along with the enforced anti-gravitational deceleration prompted by the centre of the star cluster. We could arrive at the black hole, with thick, relatively firm in consistency degenerated matter around it, at the centre of the star cluster (Image 6).

Image 6: 'Omega Centauri' Image credit: NASA/JPL-Caltech/UCLA

The Space-Time imbalance and the associated Mass-Energy and Info-Energy imbalance are reflected in the specific geometry of the multimodal systems in our dynamic world.

The two different representations of the 'o' time-point, existing in 2-dimensional time settings, confirm the coexistence of the qualities of a toroid along with the qualities of a globe in the multimodal systems. The difference is reflected in the geometry of the systems. The difference in the Space-Time and energy structure of the radiating and non-radiating system is described in my books 'Space and Time' and 'Energy of Existence'. In this book, our attention is concentrated on the black holes.

The geometry of the radiating systems may be represented as the degeneration of a toroid in our dynamic world. This historical development of the planets, stars and galaxies tangled by gravity is the basis of the Space-Time and Mass-Energy imbalance reflected in gravity on the surface of the high energy massive radiating systems and antigravity at their centres.

The geometry of the non-radiating systems, holding black holes, may be represented as the development of a toroid in our dynamic world in the future. The coexistence of the qualities of the toroid with the qualities of a globe is the basis of the Space-Time and Mass-Energy imbalance reflected in antigravity on the surface of the Space-Rising non-radiating systems and gravity at their centres.

The specific toroid-globe geometry and the associated uneven distribution of mass are the cause of the particular geometry of a multimodal system in our dynamic world.

The Space-Time imbalance, supported by the Mass-Energy and Info-Energy imbalance, specific geometry of the systems, and the balancing forces, acting between dissimilar dimensional layers of the system, are reflected in the specific rotation of these systems about their axis of rotation in our dynamic world and the associated rotation of the dynamic representations of background radiation.

The projection of the different representations of the 'o' time-point, existing in the 2-dimensional time settings, into the volume of an unbalanced Space-Rising non-radiating system, such as a system holding the central black hole or the self-rising vacuum, would draw the particular region of the Space-Time and Info-Energy imbalance with the qualities of intense gravity at the centre of the system.

The volumes, energy, and matter, absorbed via the transmitting black holes, are being processed and transformed into the forms of space, energy, and matter of the particular Space-Time region, which this black hole develops. The black hole evaporates if the multimodal system is balanced and the resultant force equals zero.

The Space-Time imbalance, supported by the Mass-Energy and Info-Energy imbalance, specific geometry of the Space-Rising non-radiating systems, and the balancing forces, acting between dissimilar dimensional layers of the system, are reflected in the specific rotation of these systems about their axis of rotation in our dynamic world and the associated rotation of background radiation that must be detectable.

The rotation of the centre of the non-radiating system is currently detectable - it influences the rotation of the matter before it falls onto a black hole. Other aspects of the non-radiating and toroidal objects and systems, developing the 2-dimensional potential space and accumulating tremendous potential energy resources (as Feynman and Wheeler mentioned), are currently unknown to us. Their latency is associated with the limits of our perception of the potential space, time, and the associated potential Info-Energy.

The Equivalence principle

The Equivalence principle of General Relativity states that 'at any point of Space-Time the effects of a gravitational field cannot be experimentally distinguished from those due to an accelerated frame of reference' (Oxford dictionary).

We, supporting the Equivalence principle of General Relativity, view the mentioned gravitational field as the Space-Time imbalance, such as the excessive time and deficit of space supported by the associated Info-Energy and Mass-Energy imbalance, which humans perceive as gravity. At any point of Space-Time, the effects of the specific Space-Time imbalance, such as the excessive time and deficit of space, supported by the Info-Energy and Mass-Energy imbalance and reflected in gravity, cannot be experimentally distinguished from those due to an accelerated frame of reference.

We, developing the Equivalence principle of General Relativity, view antigravity as the specific Space-Time imbalance, such as the excessive space and deficit of time supported by the Info-Energy and Mass-Energy imbalance. At any point of Space-Time, the effects of the specific Space-Time imbalance, such as the excessive space and deficit of time, supported by the Info-Energy and Mass-Energy imbalance and reflected in antigravity, cannot be experimentally distinguished from those due to a decelerated frame of reference.

Central Transmitting Black Holes

A transmitting black hole, joining two or more unbalanced objects or systems together, builds the system with two or more centres.

The existence of the black holes is a consequence of the system's Space-Time, Info-Energy, and Mass-Energy imbalance. An unbalanced system might develop the transmitting black holes equilibrating the system.

The systems, involved in a black hole construction, might be far away from each other in space. They are connected within the 2-dimensional Space-Time region and central points of symmetry of each Matrix. The central black hole connects the central points of symmetry of the associated Matrixes and the shared 2-dimensional Space-Time areas of the Matrixes connection.

The Arrows of Time are not harmonically balanced within these systems, and the system's Space-Time imbalance and related Info-Energy and Mass-Energy imbalance are reflected in the signs of gravity or antigravity within the central areas of the system and its central black hole. The central black hole evaporates if the system is balanced in its Space-Time, Info-Energy, and Mass-Energy and the resultant Matrix force equals zero.

Black holes demonstrate the Space-Time and Mass-Energy bonds with the maternal systems. Black Holes mechanics, resulting in the energy transformation and transmission, is determined by the type of a black hole.

Space-Rising black holes

The Space-Time imbalance and related Mass-Energy and Info-Energy imbalance, reflected in the excessive time and potential space - the inflated and undetectable 2-dimensional space, supported by the potential energy, along with the deficit of the actual volume, mass, kinetic energy and other energy representations acting in the volume of the system in the period of time now occurring, are mainly associated with the central areas of the Space-Rising non-radiating system and reflected in gravity.

Balancing transformations, resulting in the generation of space, kinetic energy, and mass, are reflected in gravity propagation prompted by the central areas of the Space-Rising non-radiating system. If the balancing transformations are virtually ineffective, the transmitting black hole may be formed at the centre of the system. The projection of the different representations of the '0' time-point, existing in 2-dimensional time settings, into the volume of the unbalanced non-radiating system will draw the area of the black hole's location within the system.

The black hole transmits space, matter, and energy from the outer space to the central areas of the system holding this black hole. It develops the volume and mass of the absorbing maternal system and consequently supports its growth.

The unbalanced absorbing system 'breathes in' and receives the volumes and energy via the direct Space-Time and Info-Energy transfer between Matrixes of the associated systems forming the transmitting black hole. The large unbalanced absorbing non-radiating system, expending in space and holding the black hole, utilises volumes, energy, and associated information, received via the black holes from the objects and systems existing on the outside.

Two connected Space-Rising non-radiating systems might build a Space-Rising black hole (SRBH) by contracting their connected Spaces of Time and central points of symmetry of each Matrix. The absorbing non-radiating objects or systems, involved in a black hole construction, might be distant in space. Two SRMs are connected within the 2-dimensional Space-Time areas and the central points of symmetry (Figure 3).

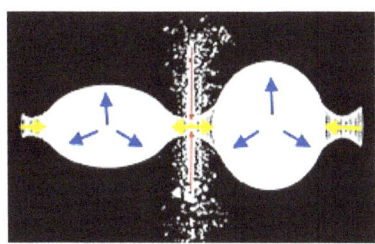

Figure 3: SRBH and two SRMs

Matter, energy, and associated information are drawn via a black hole from outer space into the central areas of each Space-Rising system and transformed into the latent Info-Energy of the 2-dimensional grid of each Matrix.

A Space-Rising black hole (SRBH) may be built by the Space-Rising non-radiating system and the high energy massive radiating system (Figure 4).

Space-Rising black holes (SRBHs) may be built by the connected Matrixes of different types, if a large and 'hungry', lacking in energy SRM of a non-radiating system is connected with a weak TRM of a high energy massive radiating system. The Space-Rising Matrix (SRM) is connected to the Time-Rising Matrix (TRM) within the 2-dimensional Space-Time areas and the central points of symmetry (Figure 4).

The central points of symmetry form a transmitting black hole supported by the connected Spaces of Time of the Matrixes.

Matter, energy and associated information are drawn from the centre of the radiating system, such as a star or a galaxy tangled by gravity, via the central points of the Matrix symmetry of these connected multimodal structures and transformed into the latent energy of the 2-dimensional grids of the TRM and connected 'hungry' SRM (Figure 4).

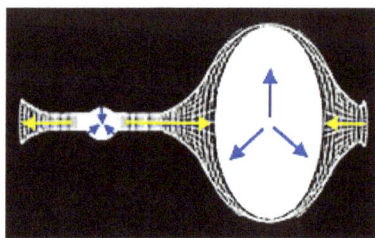

Figure 4: SRBH, TRM and SRM

If the high energy massive radiating system has a large amount of total energy, these two Matrixes (TRM and SRM) may coexist for a significant period of time, being integrated into a system with two centres that we can see as a binary system, such as the star and black hole, two bounded stars or galaxies.

The objects and systems, involved in a binary system or a multi-system construction, might be distant in space. They are connected within the 2-dimensional Space-Time region and '0' time-points of each Matrix. The development of the system is determined by the energy resources of the connected Matrixes.

Radiating black holes

A Time-Rising, or radiating, black hole (TRBH) may be built by two connected high energy massive radiating systems. Two TRMs are connected within the 2-dimensional Space-Time areas and the central points of symmetry.

If the Space-Rising system has reached the lower limits for the system's existence, such as Planck length, Planck time, and Planck energy, it will be reversed.

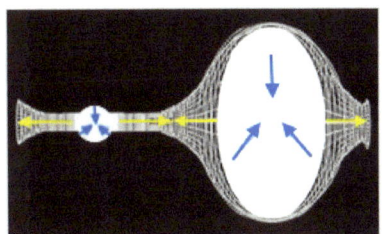

Figure 5: TRBH and Two TRMs

If the high energy massive radiating system, such as a star or a galaxy tangled by gravity, is connected with the reversed Space-Rising system, a large amount of energy and associated information of the reversed SRM is blocked by the connected TRM of the high energy massive radiating system.

The reversed unbalanced system with the predominant degenerated and reduced space and mass within the system 'breathes out', or ejects the degenerated space and energy from the centre of the system via the direct Space-Time and Info-Energy transfer between Matrixes of the associated systems forming the transmitting black hole.

The excessive masses and energy of these two connected systems, building a high degree compression upon their volumes at the centres of their Matrixes, can damage the Info-Energy grid at the centre of the weaker system, and we can detect radiation, often microwave radiation, sometimes long before we observe the jets of radiating energy in the area. The radiating black hole might be built in the damaged area following the supernova release into our dynamic world (Figure 6).

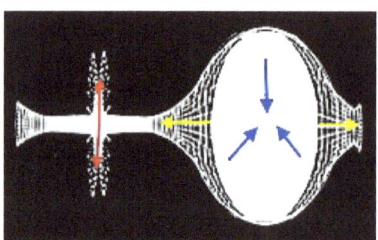

Figure 6: TRBH releasing the supernova

If the high energy massive radiating system has a large amount of energy, the excessive masses and energy of these two connected systems build a high degree compression upon the Info-Energy grid in the area of their connection (Figure 7).

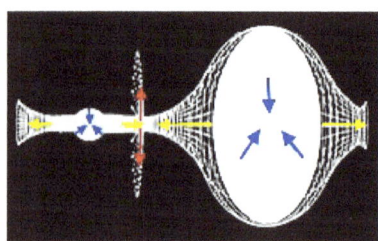

Figure 7: TRBH with jets of energy

The radiating black hole, arranged in the damaged area of the 2-dimensional grid, ejects the excessive potential space and potential energy, along with the degenerated volumes, kinetic energy, and masses, and jets of radiating energy, with the following release and reorganisation of matter and energy and creation of new high energy massive radiating and non-radiating objects and systems in our dynamic world. We can detect this radiating black hole if the 'o' time-point centre - the point 'now', is built in the damaged area.

If the intermediate centre with 'o' time-point characteristics is built between two connected Matrixes, it will be the weakest place in the system and energy may be released into the same region of our dynamic world where it is drawn from.

The Hawking radiation might be a sign of the destruction of the Info-Energy grid at the intermediate centre, which is built between two Matrixes. Similarly, it can be a sign of the destruction of the Info-Energy grid at the centre of the weaker Matrix as a consequence of the tremendous excessive Info-Energy between two Matrixes.

A black hole 'evaporates' if the SRM and the associated system - the associated region of outer space, are balanced in Space-Time, Info-Energy, and Mass-Energy, the Matrix forces are balanced, and the resultant force of the Matrix equals zero.

Surface-orientated Black Holes

The contact path, equilibrating the high energy massive radiating system, such as the Sun or our planet, might be developed by the surface of the system, if the anti-gravitational processes, associated with the degeneration and reduction of space and mass prompted by the centre of the system, would dominate the gravitational processes prompted by the system's peripheral regions. The surface orientated black holes would transmit volumes, matter, and energy to the surface of the unbalanced high energy massive radiating system holding these black holes.

The ring galaxies with external accretion, qualities of gravity prompted by the periphery of the galaxy, and anti-gravitational deceleration prompted by the galactic centre, would be an excellent example of the system developing the surface orientated black holes. These black holes with the qualities of gravity would be transmitting volumes, matter, and energy from the outer space to the surface of the unbalanced galaxy holding these black holes.

The similar contact path, equilibrating the high energy massive radiating system, would be a surface orientated black hole with the qualities of antigravity and anti-gravitational deceleration if the gravitational processes, associated with the concentration and solidification of space and mass prompted by the periphery of the system, would dominate the anti-gravitational processes prompted by the system's centre.

Radiating black holes release energy and matter under the influence of the anti-gravitational processes prompted by the centre of the system. The 2-dimensional grid vector-force of pressure, generating anti-gravitational processes at the centre of the system, can project its influence via dynamic representations of background radiation on to the surface of the system with possible creation of the radiating black holes on the surface of the system. The influence, projected on the surface of the system, is the strongest at the equator, partially neutralising a degree of gravity. We can expect these radiating black holes to be formed in the equatorial regions of the system.

The contact path, equilibrating the Space-Rising non-radiating system, such as our Universe, might be developed by the surface of the system, if the gravitational processes, associated with the concentration and solidification of space and mass prompted by the centre of the system, would dominate the anti-gravitational processes prompted by the system's periphery. The radiating black holes would release excessive energy and matter under the influence of the anti-gravitational processes prompted by the periphery of the system. The 2-dimensional grid vector-force of pressure, generating anti-gravitational processes within the system, projects its influence via dynamic representations of background radiation.

The similar contact path, equilibrating the Space-Rising non-radiating system, would be a surface orientated black hole with the qualities of gravity and gravitational acceleration if the anti-gravitational processes, associated with the degeneration and reduction of space and mass prompted by the periphery the system, would dominate the gravitational processes prompted by the system's centre. The surface orientated black holes would transmit volumes, matter, and energy to the surface of the unbalanced non-radiating system holding these black holes.

The continuing space and mass degeneration and reduction, prompted by the peripheral areas of the unbalanced absorbing system holding a central black hole, might dominate the processes of space and mass concentration and consolidation within the system, leading to the development of the toroidal form of the system in the fundamental 2-dimensional Space-Time - the potential space underlying our dynamic world.

Should we travel through the non-radiating system towards its peripheral areas, we would experience the inflation and dilation of the 'empty' space on the way as a result of space and matter degeneration and reduction, along with the enforced anti-gravitational deceleration prompted by the periphery of the system.

We could arrive at the radiating black holes delivering volumes and masses from outer space to the periphery of the system.

Alternatively, the continuing space and mass concentration and consolidation, prompted by the centre of the non-radiating system, might dominate the processes of space and mass degeneration and reduction within the system. In this case, we would arrive at the 'gravitating' black hole ejecting volumes and masses from the peripheral areas of the system into the outer space.

Image 7: 'Tale of two black holes' Image credit: NASA/JPL-Caltech/GSFC

'Before NuSTAR, astronomers knew that the each of the two galaxies in Arp 299 held a supermassive black hole at its heart, but they weren't sure if one or both were actively chomping on gas in a process called accretion. The new high-energy X-ray data reveal that the supermassive black hole in the galaxy on the right is indeed the hungry one, releasing energetic X-rays as it consumes gas.' NASA

Binary Systems of Black Holes

Binary systems of black holes reflect the process of integration of the large Space-Rising non-radiating systems holding these black holes.

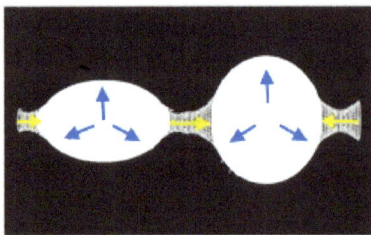

Figure 8: Integration of SRBHs

Technically, a binary system of black holes is built by the connected SRMs of the large Space-Rising non-radiating systems by contracting their connected Spaces of Time and central points of symmetry of each Matrix. The systems, involved in a binary system, might be distant in space.

The first stages of integration of the black holes may be compared with a whirlpool in a river. Rotating masses of matter gather speed in conflicting currents of the rotating black holes.

Spaces, energy, and masses are usually drawn from outer space via the black holes into the central areas of each Space-Rising system (Figure 3) and transformed into the potential Info-Energy of the 2-dimensional grid of each Matrix until the process of integration of the black holes and their maternal systems is completed. The development of the binary system is determined by the energy resources of the connected SRMs.

After the completion of the integration, spaces, energy, and matter are usually drawn from outer space via the integrated supermassive black hole into the integrated Space-Time region that this black hole develops. In our dynamic world, the researchers can often observe only matter that is drawn into the supermassive black holes.

Image 8: Two black holes on way to becoming one (Artist's Concept) Image credit: NASA

'Two black holes are entwined in a gravitational tango in this artist's conception. Supermassive black holes at the hearts of galaxies are thought to form through the merging of smaller, yet still massive black holes, such as the ones depicted here. NASA's Wide-field Infrared Survey Explorer, or WISE, helped lead astronomers to what appears to be a new example of a dancing black hole duo. Called WISE J233237.05-505643.5, the suspected black hole merger is located about 3.8 billion light-years from Earth...' NASA

*Image 9: 'Bursting with stars and black holes (Artist Concept)'
Image credit: NASA/JPL-Caltech*

'A growing black hole, called a quasar, can be seen at the center of a faraway galaxy in this artist's concept. Astronomers using NASA's Spitzer and Chandra space telescopes discovered swarms of similar quasars hiding in dusty galaxies in the distant universe. The quasar is the orange object at the center of the large, irregular-shaped galaxy. It consists of a dusty, doughnut-shaped cloud of gas and dust that feeds a central supermassive black hole. As the black hole feeds, the gas and dust heat up and spray out X-rays, as illustrated by the white rays... While some quasars are easy to detect because they are oriented in such a way that their X-rays point toward Earth, others are oriented with their surrounding doughnut-clouds blocking the X-rays from our point of view. In addition, dust and gas in the galaxy itself can block the X-rays.' NASA

Gravity of the Black Holes

The existence of black holes at the centres of the large Space-Rising systems is a consequence of the Space-Time imbalance and the associated Mass-Energy and Info-Energy imbalance, reflected in gravity. The transmitting black hole 'gravitates' and absorbs volumes, matter, and energy from the outer space if the Space-Time imbalance and the associated Info-Energy and Mass-Energy imbalance at the centre of the maternal non-radiating system reflect the excessive time and time-associated latent Info-Energy, along with the deficit of the actual space, mass, kinetic energy, and other energy representations acting in the volume of this system in the period of time now occurring.

The primary process in the central areas of the spacious non-radiating systems is the transformation of the potential Info-Energy of the 2-dimensional grid into the masses, kinetic energy and other energy representations, acting in the volume of the systems and the associated transformation of time and potential space - the inflated and undetectable space of the 2-dimensional grid, into the volume of the system, which this Matrix develops. Space-Time transformations are the function of energy. This process is reflected in the specific space and mass consolidation and qualities of intense gravity at the centres of Space-Rising non-radiating systems.

Gravity and gravitational acceleration, developed by the black hole, reflect the Space-Time, Mass-Energy, and Info-Energy imbalance and the balancing reorganisation of space, time, matter, and energy at the centre of the maternal Space-Rising system.

The non-radiating system might receive Space-Time and Info-Energy transfer directly and indirectly.

The maternal non-radiating system, such as a toroidal system, holding the central black hole, receives the volumes, matter, and energy directly from the central regions of the high energy massive radiating system, such as a white dwarf, via the transmitting black hole - the direct Space-Time and Info-Energy transfer between Matrixes of these associated systems within the binary system.

Alternatively, the non-radiating system might receive Space-Time and Info-Energy transfer indirectly via the binary system's shared area of the 2-dimensional grid that is arranged by the 2-dimensional representations of background radiation such as cosmic background radiation. Space, time, masses, kinetic energy, and other energy representations of the objects and systems are drawn via the black hole into the 2-dimensional Matrix grid of the system, holding this black hole, and transformed into time or potential space of this system. The 2-dimensional grid acts as a membrane transforming the absorbed volumes, matter, and energy into the potential space and space-associated potential energy of the shared 2-dimensional Space-Time.

Black holes gravitational horizon is the gravitational horizon of the system, forming a black hole.

The gravitational horizon of the system, forming a black hole, is the system's Info-Energy grid, which is the boundary of the system at the centre of its multimodal SRM and internal dynamic Space-Time and Info-Energy skeleton of this system in our dynamic world. The Matrix grid, arranged by the 2-dimensional representations of background radiation, conducts the Space-Time, Info-Energy, and Mass-Energy transformations reflected in gravity.

Transformations associated with gravity:

A large amount of time is required in order to get a small amount of space proportional to time with the application of the Coefficient of Transformation $[c]^2$ if the Mass-Energy balance of the object is unchanged.

A large amount of energy is required in order to get a small amount of mass proportional to energy with the application of the Coefficient of Transformation $[c]^2$ if the Space-Time balance of the object is unchanged.

A large amount of latent, or potential, Info-Energy is required in order to get a small amount of actual Info-Energy proportional to latent, or potential, Info-Energy with the application of the Coefficient of Transformation $[c]^2$ if the Space-Time balance of the object is unchanged.

The Coefficient of Transformation $[c]^2$ applies to the decisions associated with the Space-Time, Mass-Energy and Info-Energy Transformations reflected in gravity.

The Coefficient of Transformation $[c]^2$ equals the squared numerical value of the speed of electromagnetic waves propagation in a vacuum.

In our dynamic world the Space-Time, Mass-Energy and Info-Energy transformations, including those associated with gravity, propagate with the speed of light and are carried by background radiation.

The speed of light is a CODATA Fundamental Physical Constant. The value of the speed of light in a vacuum (c) is 299 792 458 m s-1 (meter per second).

The development of the binary system is determined by the structure of the Space-Time and Info-Energy resources of the associated multimodal systems.

Image 10: 'Stellar rubble may be planetary building blocks (Artist Concept)' Image credit: NASA/JPL-Caltech

'This artist's concept depicts a type of dead star called a pulsar and the surrounding disk of rubble discovered by NASA's Spitzer Space Telescope. The pulsar, called 4U 0142+61, was once a massive star until about 100,000 years ago when it blew up in a supernova explosion and scattered dusty debris into space. Some of that debris was captured into what astronomers refer to as a "fallback disk," now circling the remaining stellar core, or pulsar... The pulsar observed by Spitzer, called 4U 0142+61, is 13,000 light-years away in the northern constellation Cassiopeia. Its disk orbits about 1 million miles (1.6 million kilometers) away from it...' NASA

Reverse Of The Absorbing System

The large Space-Rising non-radiating systems, reaching the limits of the system's existence, will be reversed into the high energy massive radiating systems (Figure 9).

The continuing space and mass generation, prompted by the centre of a spacious non-radiating system, might dominate the space and mass degeneration, prompted by the system's peripheral areas. It would result in the rising Space-Time and Info-Energy imbalance under the influence of the system's vector-force of resistance developing the system's Space of the Current Time and associated actual Info-Energy represented in mass and kinetic energy acting in the volume of the system in the period of time now occurring. The Space-Time and Info-Energy imbalance is reflected in the intense gravity prompted by the centre of the system. The excessive actual space and masses can damage the system's centre and create the central black hole.

The process coexists with the rising deficit of the potential space, time, and the associated potential Info-Energy resources, damage of the 2-dimensional grid, and creation of the transmitting black holes within the system's 2-dimensional grid.

The following limitation of the potential Space-Time and Info-Energy resources will reverse the Space-Time and Info-Energy flow within the system's Spaces of Time and the system's black holes. It will bring about the generation of spaces and masses, along with the gravitational processes prompted by the periphery of the system and the development of inflation, anti-gravitational deceleration, and reduction of volumes, masses, kinetic energy, and other energy representations prompted by the system's centre. The non-radiating system will be reversed into the high energy massive radiating system. Accordingly, the system's SRM will be reversed into the TRM (Figure 9).

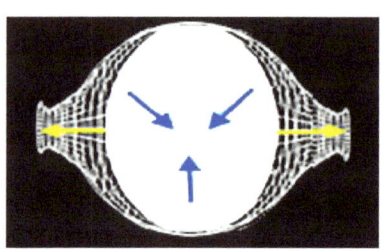

Figure 9: Reverse of the SRM

Alternatively, the continuing space, kinetic energy, and mass degeneration and reduction, along with anti-gravitational deceleration prompted by the periphery of the non-radiating system, might dominate the processes of the space, kinetic energy, and mass generation reflected in gravitational acceleration and prompted by the system's centre. It would lead to the development of the toroidal form of the system in the 2-dimensional Space-Time and the intense antigravity prompted by the periphery of the system. The deficit of the actual space, masses, and kinetic energy could damage the centre of the system and form the central black hole.

The following limitation of the actual Space-Time and Info-Energy resources of the system will reverse the direction of the Space-Time and Info-Energy flow (Figure 9) and initiate deflation, gravitational acceleration, and generation of volumes, masses, and kinetic energy prompted by the peripheral regions of the system. The reverse of the Space-Time and Info-Energy flow within the system will reverse the Space-Time and Info-Energy flow in the system's black hole. The central black hole will transmit the 2-dimensional Space-Time and associated potential Info-Energy, along with the newly created space and matter from the system's centre into the outer space.

The Space-Rising non-radiating system will be reversed into the high energy massive radiating system. Accordingly, the system's SRM will be reversed into the TRM (Figure 9).

The reverse of the direction of the Space-Time and Info-Energy flow will change the properties of the system. The reversed non-radiating systems, passing the period of their development into the high energy massive radiating systems, create supernovae in our dynamic world.

Our young Universe, rising in size, continues progressing as the Space-Rising non-radiating system. If our Universe stretches to the state of the 2-dimensional toroid and reaches the limits actuating the reverse, it will be reversed by the Matrix forces.

The reverse of the direction of the Space-Time and Info-Energy flow will initiate deflation, gravitational acceleration, and generation of volumes, masses, kinetic energy, and other energy representations prompted by the peripheral regions of the Universe.

Accordingly, the Universe would be reversed into the high energy massive radiating system and could create the supernova in the associated modality of the Grand Universe.

The Big Bang theory as a cosmological model of the Universe is struggling with the starting point of the Universe' existence in time, space, mass, and energy. It also has difficulties with the Mass-Energy Conservation Law.

It does not have to be that way. Nothing is coming from nothing and going to nothing.

'By laying down the relativity postulate from the outset, sufficient means have been created for deducing henceforth the complete series of laws of mechanics from the principle of conservation of energy (and statements concerning the form of the energy) alone.' [Minkowski Hermann, The Fundamental Equations for Electromagnetic Processes in Moving Bodies. Appendix (1908)]

We must not put the principle of conservation of energy aside and ignore the basics of the quantum theory.

The '0' Space-Time and '0' energy point as the start or the end of the Universe will never be found or proven because they never exist. They are not achievable in our dynamic world of matter and related exclusively to the human perception of time as a point 'now'. Planck length, Planck time, and Planck energy reflect the minimal Space-Time and Info-Energy imbalance in the existing objects and systems as the necessary deviations from the '0' Space-Time point and '0' energy point.

The speed of light in vacuum and the Planck units, based on the calculations using the speed of light in vacuum as the fundamentals proven by experiments - Planck length, Planck time, and Planck energy, provide the upper and lower limits for the existence of the Universe as we measure and sense it.

If a system has reached the limits actuating the reverse, the Space-Time and Info-Energy flow within the system is being reversed following the Laws of Space-Time, Mass-Energy, and Info-Energy Conservation, Reversibility, Limitation, Transformation, and Symmetry.

Reverse of the Black Holes

The Space-Rising non-radiating system, keeping a black hole and receiving volumes, masses, and energy from the outer space, is a part of a binary system or a multi-system. It is usually associated with the high energy massive radiating system it feeds from. The reverse of the direction of the Space-Time and Info-Energy flow within the system changes the direction of the Space-Time and Info-Energy flow in the central black hole to the opposite. The 2-dimensional grid vector-force of pressure, building a high degree pressure upon the structure of the central black hole, conducts the reverse of Space-Time and Info-Energy flow within the central black hole.

We can take the oldest discovered supernova as an example. The region of low-density, surrounding the supernova, was created before the explosion of the supernova by the black hole that was sending matter, energy, and volume from the central areas of the white dwarf and surrounding areas (the case of the multi-system involved in the construction of the black holes) into the distant Space-Time region - the Space-Rising non-radiating system. The black hole fed this remote absorbing system. The volumes, matter, and different forms of energy had been absorbed via the 'gravitating' black hole and transmitted via SRM into the centre of the remote, possibly toroidal, system.

The reverse of the distant absorbing system initiated the reverse of the Space-Time and Info-Energy flow within the existed black hole. Accordingly, the 'radiating' black hole ejected the volumes, matter, and energy from the centre of the reversed system into the central areas of the white dwarf creating the oldest discovered supernova.

The following dynamic balance within the system, holding a black hole, happens when a process and its reverse are occurring at equal rates. The dynamic balance within the system leads to the dynamic balance within the central areas of the system and the radiating black hole 'evaporates'. The radiating black hole 'evaporates' if the Space-Time, Info-Energy, and Mass-Energy of the system are balanced, Matrix forces are balanced, and the resultant force of the Matrix equals zero.

Image 11: 'Blazar (Artist Concept)' Image credit: NASA/JPL-Caltech/GSFC

'Black-hole-powered galaxies called blazars are the most common sources detected by NASA's Fermi Gamma-ray Space Telescope. As matter falls toward the supermassive black hole at the galaxy's center, some of it is accelerated outward at nearly the speed of light along jets pointed in opposite directions. When one of the jets happens to be aimed in the direction of Earth, as illustrated here, the galaxy appears especially bright and is classified as a blazar.' NASA

Antigravity of the Black Holes

Space and mass creation is prompted by the central areas of the Space-Rising non-radiating objects and systems. It coexists in a dynamic balance with space and mass degeneration and reduction prompted by their periphery. The generated space, energy, and matter, all decline throughout the system into the various forms of the degenerated space, non-radiating matter, and energy. They are detectable in the regions of inflation within the Universe.

The ring galaxies with external accretion and cohesion of matter under the influence of gravitation, prompted by the periphery of the galaxy, and antigravity, prompted by the galactic centre, would be an excellent example of the system developing the transmitting black holes. Should we travel through these type of the ring galaxies towards the galactic centre, we would experience resistance of the 'empty' space on the way as a result of space dilation (the inflated space) or extension with the faster speed of time, and enforced anti-gravitational deceleration prompted by the galactic centre.

In the systems trending towards antigravity, 'normal' matter, energy, and space degenerate into the inflated space, degenerated non-radiating matter and energy, and pass into the reduction - the volumes, matter, and energy are being transformed into the 2-dimensional Space-Time supported by the latent, or potential, Info-Energy thus shifting the power from the centre to the periphery of the system, and finally transforming the Space-Rising non-radiating system into the toroidal system.

Space, energy, and matter, generated under the influence of gravity, degenerate through the system under the influence of antigravity creating inflated space and degenerated matter and energy - dark matter and dark energy. Antigravity propagation throughout the unbalanced spacious and toroidal systems bring about the degeneration and reduction of space and matter, which are being transformed into the potential space and the associated potential Info-Energy.

The Space-Time, Info-Energy, and Mass-Energy imbalance, reflected in antigravity, can be responsible for the creation of the transmitting black holes on the periphery and in the central regions of the large Space-Rising non-radiating systems and at the centres of the toroidal system reversed into the radiating systems. The transmitting black holes carry space, matter, and energy away from the maternal systems into the outer space, for example, the centre of the associated system.

Antigravity and anti-gravitational deceleration, developed by the black hole, reflect the Space-Time, Mass-Energy, and Info-Energy imbalance reflected in antigravity and the balancing reorganisation of space, time, matter, and energy at the centre of the maternal system or its peripheral areas.

'Astronomers don't know why the hidden black holes would have larger halos of dark matter but are intrigued by the surprising finding and are investigating further' (NASA).

Perhaps, we should pay attention to the black holes, balancing the anti-gravitational processes within the 2-dimensional Info-Energy grid in the peripheral areas of the large unbalanced Space-Rising and toroidal non-radiating objects and systems with an extensive collection of degenerated space, energy, and matter, including dark matter and dark energy.

On the other hand, the reversed toroidal system will change the direction of Space-Time and Info-Energy flow at the central black hole to the opposite.

The central black hole 'radiates' taking the excessive potential space, energy and related information away from the system and balancing the system. Similarly, gases flow from a region of higher pressure to one of lower pressure.

The specific Space-Time imbalance, such as the excessive potential space and time deficit, and associated Info-Energy and Mass-Energy imbalance within the central areas of the reversed large spacious and toroidal 2-dimensional systems, holding radiating black holes, are reflected in antigravity. The dominant influence of the 2-dimensional grid vector-force of pressure is intense in the central areas of the reversed toroidal systems. This continuous physical force of pressure is the measure of the excessive latent, or potential, Info-Energy of the 2-dimensional grid. The excessive Info-Energy, associated with the potential space, creates a high degree pressure in the central area of the reversed unbalanced system holding the radiating black hole.

The Space-Time transformations and associated Info-Energy and Mass-Energy transformations, reflected in anti-gravity transmission, are most intense in the central areas of the reversed system and within the central radiating black hole.

The reversed unbalanced spacious and toroidal systems, holding radiating black holes at the central areas, use black holes as built-in tunnels, ejecting from the system potential and actual space, energy, and new-formed matter creating supernova in our dynamic world.

'While searching the skies for black holes using NASA's Spitzer Space Telescope, astronomers discovered a giant supernova that was smothered in its own dust. In this artist's rendering, an outer shell of gas and dust - which erupted from the star hundreds of years ago - obscures the supernova within. This event in a distant galaxy hints at one possible future for the brightest star system in our own Milky Way.' NASA

The anti-gravitational horizon is the system's 2-dimensional Info-Energy grid, processing Space-Time and Info-Energy transformations. Dynamic representations of background radiation arrange, support, transport, and transmit the balancing Space-Time, Info-Energy, and Mass-Energy transformations, including those reflected in anti-gravitational deceleration, through our dynamic world.

The Coefficient of Transformation $[c]^2$ applies to the decisions associated with the Space-Time, Mass-Energy, and Info-Energy transformations reflected in antigravity propagation.

The Theory of Matrix reflects Space-Time, Info-Energy, and Mass-Energy transformations, associated with antigravity, as follow:

A small amount of space is required in order to get a large amount of time proportional to space with the application of the Coefficient of Transformation $[c]^2$ if the Mass-Energy balance of the object is unchanged.

A small amount of mass is required in order to get a large amount of energy proportional to mass with the application of the Coefficient of Transformation $[c]^2$ if the Space-Time balance of the object is unchanged.

A small amount of actual Info-Energy is required in order to get a large amount of latent, or potential, Info-Energy proportional to actual Info-Energy with the application of the Coefficient of Transformation $[c]^2$ if the Space-Time balance of the object is unchanged. In our dynamic world, the Space-Time, Info-Energy, and Mass-Energy transformations, associated with antigravity, propagate with the speed of light and are carried by background radiation. Please see my book 'Space and Time' for details associated with the Space-Time Coefficient.

Dynamics preserve the Laws of Space-Time, Mass-Energy, and Info-Energy Conservation, Reversibility, Limitation, Transformation, and Symmetry.

Rotation of the Black Holes

The primary rotation of a black hole depends upon the Space-Time imbalance and the associated Mass-Energy and Info-Energy imbalance in the Space-Time region of the black hole location, for example, the Space-Time region at the centres of the unbalanced systems, in the peripheral areas of the unbalanced systems, between dissimilar dimensional layers of the system, within the 2-dimensional Space-Time area of the systems' connection or the intermediate centre with '0' time-point characteristics, which are built by the connected multimodal systems involved in the black hole construction.

The rotation of the black hole might be determined by the dominance of the Info-Energy grid vector-force of pressure or the system's vector-force of resistance.

Four different types of rotation of a black hole depend on the black hole location and the type of the Space-Time and Info-Energy imbalance in the area. The location of the black hole at the centre or on the periphery of the system might determine its rotation. The rotation of the black hole might also be determined by gravity or antigravity in the area.

The Space-Time, Info-Energy and Mass-Energy imbalance and specific geometry of the multimodal systems, along with the associated influence of the Matrix forces, are reflected in the specific rotation of the volume of these multimodal systems about their Space-Time axis, along with the relative specific rotation of the Matrix grid, and impacts the rotation of the black hole. The relative specific rotation of the Matrix grid is reflected in the rotation of cosmic background radiation. The specific rotation of the background radiation might influence the rotation of the central black hole.

The secondary rotation of the black holes depends upon the rotation of the objects and systems involved in the black hole construction, the rotation of the 2-dimensional grid involved in the black hole construction and associated rotation of dynamic representations of background radiation, such as cosmic background radiation.

The secondary rotation of the black holes is also influenced by the properties of the black hole, such as its size, and the amount, direction and rotation of the volumes, matter and energy flow moving through the black hole.

On the other hand, the black hole's size and rotation influence the amount, direction and rotation of the volumes, matter, and energy flow moving into the black hole.

The reverse of an unbalanced multimodal system affects the specific rotation of this system, rotation of the background radiation, and the associated rotation of the black hole.

The newly established Space-Time and Info-Energy imbalance can change the direction of rotation about the system's Space-Time axis to the opposite and impact the rotation of the reversed system in our dynamic world.

The rotation of the reversed system about its axis of rotation would reflect a new direction of its Space-Time axis and, accordingly, a new axis of rotation in our dynamic world. It would impact on the gradual shift in the orientation of the system's axis of rotation.

The reverse of an unbalanced multimodal system, holding the black hole, also affects the rotation of the flow of matter, volumes, and energy, which this black hole transmits and affects the rotation of the Space-Time region associated via the black hole.

Image 12: 'Spitzer captures Messier 87' Image credit: NASA/ JPL-Caltech/IPAC/Event Horizon Telescope Collaboration

'This image from NASA's Spitzer Space Telescope shows the elliptical galaxy Messier 87 (M87), the home galaxy of the supermassive black hole recently imaged by the Event Horizon Telescope (EHT). Spitzer's infrared view shows a faint trace of a jet of material spewing to the right of the galaxy - a feature that was previously one key indicator that a supermassive black hole lived at the galaxy's center.' NASA

Afterword

The Theory of Matrix is a new theory combining elements of psychology, cosmology, and astrophysics. This theory was introduced by Dr Audrey E. Randles in her work 'The Theory of Matrix' in 2012.

Dr Randles developed the program for psychological investigations of the Universal Matrix along with the development of the Coresynthesis Psychological Model in early 1990s. There was a space of ten years between the first explanations of the results and association of these results with the characteristics mathematically introduced by Lorenz, Minkowski, and Einstein for the Light Cone.

When Dr Randles published her work, the Light Cone was considered a specific case that can be applied solely to the flash of light. She brought forward the idea of the universality of the Space-Time structure of the objects and a new vision on Space-Time physics.

Following the analysis of the parallels and variations between the Universal Matrix and Light Cone Dr Randles calls the Light Cone 'the Matrix of Light' and presents the Theory of Matrix as the relative importance for the true understanding of the multimodal world.

The Theory of Matrix introduces the new understanding of the multidimensional world with respect to multidimensional time. Time is no longer seen as another dimension of space, nor as a momentary feature of an event but as a multidimensional element in its own right. Dr Randles associates space with Actuality and time with Potentiality and Latency. Therefore, the objects are viewed as multimodal, multidimensional objects in Space-Time.

Books on Cosmology by Audrey E. Randles:
'Systems Theory in Cosmology' (2020)
'The Multimodal World' (2020)

'Black Holes and Supernovas' (2016)
'Grand Universe' (2016)
'Antigravity' (2015)
'Supernovas' (2015)
'The Primary Black Hole of the Universe' (2015)
'Energy in Cosmology' (2014)
'Gravity and Antigravity to the Point' (2014)
The Theory of Matrix series of books (2012 - 2013) includes the following books:
'Blocks of the Universe'
'Space and Time'
'Energy of Existence'
'Gravity and Rotation'
'Black Holes'
'Matrix of the Universe'
The New 2020 books on Cosmology include Kindle ebook and a paperback of the same title at Amazon's Book Store.

Content Use Policy

© Audrey Elizabeth Randles

Content may be used for any purpose without prior permission, subject to the special cases noted below.

By downloading the material, the user agrees:

1. to use a credit line in connection with the content. Unless otherwise noted in the caption information for any content and images the credit line should be

"Audrey E. Randles, 'The Theory of Matrix. Black Holes' (2013) red. 2020".

2. that we do not represent others who may claim to be authors or owners of copyright of any of the content, and make no warranties as to the quality of the content;

3. that we shall not be responsible for any loss or expenses resulting from the use of the content, and you release and hold us harmless from all liability arising from such use.

Special Cases:

This content is available for educational, journalistic, personal uses and scientific research following a scientific code of ethics.

Restrictions are placed on commercial uses. To obtain permission for commercial use, contact the copyright owner Dr Audrey E. Randles.

www.ingramcontent.com/pod-product-compliance
Lightning Source LLC
Chambersburg PA
CBHW040233220526
45473CB00001B/228